FINAL TOUCHDOWN

The Last Days of East German Air Power

by Hans Nijhuis

CONCORD
PUBLICATIONS COMPANY
CONCORD COLOR SERIES

603-609 Castle Peak Road
Kong Nam Industrial Building
10/F, B1, Tsuen Wan
New Territories, Hong Kong

We welcome authors who can help expand our range of books. If you would like to submit material, please feel free to contact us.

We are always on the look-out for new, unpublished photos for this series.
If you have photos or slides or information you feel may be useful to future volumes, please send them to us for possible future publication.
Full photo credits will be given upon publication.

ISBN 962-361-703-8
Printed in Hong Kong

The MiG-29 is able to carry six external stores and a fuel tank under the fuselage, between the two engines. (Stefan Petersen)

During taxiing and parking the Flogger's wing is fixed in its 'fast' setting of 72 deg.

All photos were taken by Hans Nijhuis, unless otherwise stated.

Many of the photos in this book can be obtained as original slides. Also available are slides of military aircraft from other European, American and Asian countries.

Send request for information to
TOP SLIDES
P.O. BOX 157
7570 AD OLDENZAAL
THE NETHERLANDS

INTRODUCTION

When on October 3, 1990 the re-unification of West- and East Germany became a fact, one of the most powerful armed forces of the Warsaw Pact was deactivated, while only four months later most units were disbanded. Only a very small part of the 700 aircraft and helicopters of the East German armed forces would be integrated in the West German system. This booklet illustrates the aircraft and helicopters of the East German Air Force, Army and Navy during the last months of their operational life within the German Democratic Republic. It also incorporates some photos of East German equipment that has found (temporarily or more permanently) employment in the Federal German Luftwaffe.

ACKNOWLEDGEMENTS

Most of the photos in this book were taken during the last two months of the existence of the German Democratic Republic. At that time nobody in the NVA knew what was going to happen with him because it was obvious that many of them would lose their jobs after the re-unification. Nevertheless all armed forces personnel I met were most helpful in assisting me with the realisation of my photographic requirements. From this place I wish to express my sincere gratitude for this fact because without their help the production of "Final Touchdown" would not have been possible. Though it is impossible to name all these individuals, I would like to name a few for the generous help and friendship I received: Cols Hempel and Menge, and Lt Cols Agata, Fichte, Knobloch, Kohler, Lahner, Lill, Sauder, Skiba, Sonntag, Stange, Syrbe and Ziegler.

Furthermore I like to thank the following persons for lending me material to use in this publication:

Harry Berger, Hendrik Jan van Broekhuizen, Stefan Petersen, Steffen Seifert and Martin Zijlstra.

List of Abbreviations

Abbrev.	German	English
ASCC	-	Air Standards Coordinating Committee
FAG	Fliegerausbildungsgeschwader	Air training wing
FDK	Flugzieldarstellungskette	Target towing flight
FLIR	-	Forward Looking Infra-Red
HAG	Hubschrauberausbildungsgeschwader	Air training wing
HSFA	Hubschrauberstaffel der Führung und Aufklärung	Command and observation helicopter squadron
JBG	Jagdbomberfliegergeschwader	Fighter Bomber Wing
JG	Jadgfliegergeschwader	Fighter Wing
KHG	Kampfhubschraubergeschwader	Combat helicopter wing
LLTV	-	Low Light Television
LSK/LV	Luftstreitkräfte/Luftverteidigung	Air Force/ Air Defence Force
MAD	-	Magnetic anomaly detector
MFG	Marinefliegergeschwader	Naval air wing
MHG	Marinehubschraubergeschwader	Naval helicopter wing
NVA	Nationale Volksarmee	Armed Forces
OKB	-	design bureau
SAR	-	Search and rescue
TAFS	Taktische Aufklärung Fliegerstaffel	Tactical air reconnaissance squadron
TAS	Transportfliegerausbildungsstaffel	Air transport training squadron
THG	Transporthubschraubergeschwader	Transport helicopter wing
TG	Transportfliegergeschwader	Air transport wing
QRA	-	Quick Reaction Alert
TS	Transportfliegerstaffel	Air transport squadron
VFK	Verbindungsflugkette	Air communications flight
VFS	Verbindungsfliegerstaffel	Air communications squadron

Air Order of Battle on October 1, 1990

Unit	Name	Base	Squadrons	Aircraft type	Number
LUFTSTREITKRÄFTE/LUFTVERTEIDIGUNG					
1. Luftverteidigungsdivision					
VFK /1	-	Cottbus	-	An-2/Z-43	2/3
JG 1	Fritz Schmenkel	Holzdorf	2	MiG-21MF	35
			1	MiG-21SPS-K/UM	21/9
JG 3	Wladimir Komarow	Preschen	2	MiG-29A/UB	20/4
			1	MiG-21MF/MF-75/UM	3/9/3
JG 8	Hermann Matern	Marxwalde	3	MiG-21bis/UM	41/8
3. Luftverteidigungsdivision					
VFK/3	-	Neubrandenburg	-	An-2/Z-43	2/2
JG 2	Juri Gagarin	Neubrandenburg	2	MiG-21M	31
			1	MiG-21SPS-K/UM	12/8
JG 9	Heinrich Rau	Peenemünde	2	MiG-23ML	28
			1	MiG-23MF/UB	9/6
FDK 33	-	Peenemünde	-	L-39V/ZO	2/2
Kommando der Front- und Transportfliegerkrafte					
JBG 37	Klement Gottwald	Drewitz	2	MiG-23BN/UB	18/3
JBG 77	Gebhard Leberecht von Blucher	Laage	2	Su-22M-4/Su-22UM-3K	22/4
TAFS 47	-	Preschen	1	MiG-21M/UM	13/3
TAFS 87	-	Drewitz	1	MiG-21M/UM	12/2
THG 34	Werner Seelenbinder	Brandenburg-Briest	2	Mi-8T/Mi-8S	17/2
TG 44	Arthur Pieck	Marxwalde	2	Mi-8S/Tu-134A *	6/1
TS 24	-	Dresden	1	An-26	12
VFS 14	-	Strausberg	-	An-2/L-410UVP/Z-43	8/4/4
Offiziershochschule der LSK/LV fur Militarflieger					
	Otto Lilienthal	Bautzen			
FAG 15	Heinz Kapelle	Rothenburg	2	MiG-21SPS/SPS-K	22/4
			2	MiG-21/U/US/UM	13/13/3
FAG 25	Leander Ratz	Bautzen	3	L-39ZO	48
HAG 35	Lambert Horn	Brandenburg-Briest	2	Mi-2/8S/T	17/11/6
TAS 45	-	Kamenz	-	An-2/L-410T/Z-43	12/8/3
ARMEEFLIEGERKRAFTE					
KHG 3	Ferdinand von Schill	Cottbus	1	Mi-24D	19
			1	Mi-8TB	16
HSFA 3	-	Cottbus	-	Mi-2/8S/9	4/1/4
KHG 5	Adolf von Lutzow	Basepohl	1	Mi-24P	12
			1	Mi-24D	20
			1	Mi-8TB	10
HSFA 5	-	Basepohl	-	Mi-2S/8S/9	4/1/4
VOLKSMARINE					
MHG 18	Kurt Barthel	Parow	1	Mi-8S/T/TB	1/2/11
			1	Mi-14BT/PL	6/10
MFG 28	Paul Wieczorek	Laage	2	Su-22M-4/Su-22UM-3K	23/4

* Also some civil registered Tu–134As, Tu–154Ms and Il–62s.

This photo of a MiG-29A shows the closed louvre air intakes over the wingroot and the chaff/flare cartridge housings in front of the vertical tailplanes.

A MiG-29A 'Fulcrum' passes the control tower of the line chief at Preschen.

'669' is a basic MiG-29A (industry designation Type 9-12A), like the ones in Soviet service. The type 9-12B is the export version that went to countries like Iraq.

The NVA Fulcrums were very low on hours at the time of re-unification. This example had flown only 141 hours until August 1990.

A modified Type 9-12A. Internal changes are the autopilot and navigation computer. External changes comprise a different operation of the air intakes, a nosewheel mudguard and modified brake parachute housing.

20 MiG-29As and four MiG-29UBs equipped the first and second squadron of Jagdfliegergeschwader 3 'Wladimir Komarow' at Preschen. JG 3 belonged to Luftverteidigunsdivision 3, together with JG 1 and JG 8.

With its 'empty' weight of 24,250 lb (11,000 kg.) and a landing speed of 135 knots (250 km/h) the MiG-29 normally uses a brake parachute to reduce speed.

Evaluation of the MiG-29's sensor and weapon capabilities showed that its performances are close to that of the F-15 and F-18 fighters. (Stefan Petersen)

The MiG-29 has a limited action radius due to its two fuel-consuming RD.33 engines. (Stefan Petersen)

To commemorate the last flight of the MiG-29 in NVA service, this aircraft received a special painting scheme.

Lt Col Fichte taxies back to Preschen's flightline after the ultimate flight of MiG-29A '604' in NVA service on 27 September 1990.

'604' was the first MiG-29A delivered to the NVA in March 1988 and it became '29-01' in the Luftwaffe designation system. Until the summer of 1988 the existence of NVA MiG-29s was kept secret from the East German population, although its arrival in East Germany drew the attention of Western pressmedia in April 1988.

Transition from NVA to Luftwaffe: MiG-29A '29-01' (ex '604'). The MiG-29 is the only NVA fighter that will see operational service in the Luftwaffe.

Nearly an empty flightline at Preschen on July 3, 1991: No MiG-21s left and MiG-29 operations were limited to only a few flights each week.

'29-14' taxies out of the Preschen flightline. Note the closed air intakes to avoid FOD.

This photo shows the air intakes in open position. Also note the absence of the nosewheel mudguard which identifies '604' as a non-modified MiG-29A (Type 9-12A). In fact '604' to '679' ('29-01' to '29-12') are all non-modified, while '684' to '786' ('29-14' to '29-21') were delivered with modifications.

27.09.90 is the date of the last MiG-29 flight in the NVA.

On October 19, 1990 the Preschen based Geschwader commenced flying activities within the Luftwaffe. It is expected that the MiG-29 unit will move to Altenburg airbase, near Leipzig, within two years, after the Soviets have handed over that base to the German authorities.

Just before his last flight in NVA service, Lt Col Fichte, one of the two MiG-29 squadron commanders poses for his aircraft that for the occasion wears the JG 3 insignia.

The large aerial on the MiG-29's tail encloses the VHF antenna, just below it the CO-69 radar warning antenna and the "Swift Rod' ILS aerial are located.

After a (peaceful) confrontation with Dutch F-16 pilots the MiG-29 pilots found out their aircraft did not require upgrading, although they personally adopted some habits in the pilot's outfit.

The air brakes are constructed around the brake parachute housing. This example is of a non-modified MiG-29A. The cover is attached to the parachute and falls with it on the runway. The cover of the modified type is attached to the parachute housing.

The cockpit of the MiG-29A has less 'glass' than that of Western fighters of the same category.

MiG-29UB '179' returns after a training mission in August 1990. Note the periscope for the backseat pilot that is used during taxiing, take-off and landing.

The MiG-29 just fits into its shelter. The rails are used to keep the aircraft on centerline when it is pushed back into the shelter.

'181' accumulated over 600 flying hours early August 1990. The MiG-29UB was designated Type 9-51 by the Mikoyan OKB.

Normal flight operations were carried out from the flightline as a matter of efficiency. Only during exercises the aircraft would operate directly from its shelter.

The turn-around time of the Fulcrum is very quick, only some 30 minutes.

All maintenance instructions on the MiG-29 are in Russian language.

Ex '185' during an overshoot. The louvre air intakes are just closing.

A MiG-29UB waiting for clearance to taxi out. Note that the periscope is in retracted position.

Col Manfred Menge, a former West German F-4 wing commander who became CO of the MiG-29 wing after the re-unification, inspects 'his' aircraft prior to a flight. This aircraft ('29-14') has English cockpit displays instead of Russian. The emergency audio warning system 'Natasja' of this specific aircraft was reprogrammed in German language and is now referred to as 'Isabella'.

By using the helmet sighted visor a MiG-29 pilot is able to react fast in air combat situations.

The rear cockpit of the MiG-29UB is not only instrumented for flight training. It also has an instrument panel to control the simulation of radar and IR targets for the forward displays.

A MiG-23BN 'Flogger H' is halted for a safety check just after landing at its home base.

This photo clearly shows the armour plates alongside the cockpit.

A MiG-23BN of the second squadron of Jagdbombefliegergeschwader 37 'Klement Gottwald'. The extended nosewheel position indicates that this photo was taken after landing.

'701' rolling to the runway of Drewitz. The last flying day of JBG 37 within the NVA was on September 25, 1990.

This aircraft is armed with launchers for unguided 57mm rockets. Drewitz maintained a QRA force of six Floggers, each armed with 96 unguided rockets.

This aircraft demonstrates the variable geometry at its 16 deg. sweep setting.

An aircraft of the first Staffel (squadron) with a centerline fueltank.

In total the NVA received 22 Flogger Hs from 1979 onwards.

As soon as the landing gear is downed, the vertical stabilizer below the tail is automatically locked in a horizontal position.

The main task of the MiG-23BN is interdiction. During its 12 years of operation JBG 37 lost four of these aircraft. Due to safety restrictions the Flogger mostly was flying en-route altitudes over 1,000 ft. The pictured aircraft is a MiG-23ML which has a radar built in its nose.

'697' in take-off position at the Drewitz runway. The pilots of JBG 37 nick-named their aircraft 'jetglider' because of its aerodynamical properties.

The 'Q' sign on the nose was an award for this aircraft's excellent state of maintenance. The maximum number of stars obtainable was 5.

'691' is running its engine up to take-off power. The MiG-23 is very noisy, especially at close range!

Returning after an early morning mission this MiG-23BN is taxiing to the flightline. Half an hour later the aircraft will take off again for its next mission.

Like most Soviet aircraft, the MiG-23 has a poor sight to the rear. Therefore a mirror has been constructed on top of the canopy. The forward view is quite good compared with the radar-equipped MiG-23MF/ML.

Cockpit interior of the MiG-23BN. Several instruments are similar to those in the MiG-21 and MiG-29.

'20-39' is the ex '690' of the first squadron. It is stored at Drewitz, like most other MiG-23BNs.

Ex '710' holds its wings in a 45 deg. position which was most suited for manoeuvring.

MiG-23ML '338' drags its chute off the runway after an evening flight.

Peenemünde based Jagdfliegergeschwader 9 was tasked with air defence and comprised two squadrons MiG-23ML and one squadron MiG-23MF.

One of four aircraft that were always on 10 min. readiness.

The runway floodlights were normally only used to light the runway during night approaches.

A pair of MiG-23MLs preparing for an evening take-off. Sometimes JG 9 Floggers met NATO aircraft in international airspace over the Baltic.

At the time of re-unification the NVA had 28 MiG-23MLs in service.

This aircraft is equipped with a practice missile.

With its wings set at 16 deg. this MiG-23 interceptor is overflying the Baltic coast. (Stefan Petersen)

Jagdfliegergeschwader 9, 'Heinrich Rau' belonged together with JG 2 'Juri Gagarin' to Luftverteidigungsdivision 3. (Stefan Petersen)

A MiG-23ML 'Flogger G" over East Germany at sunset. The wingsweep is at manoeuvring position. (Stefan Petersen)

Typical armament for the alert force consisted of two radar-guided R-23T 'Apex' and four IR guided R-60 'Aphid' missiles.

The R-23R is the IR guided version of the 'Apex' missile. The missile has a warhead of 66 lb (30kg) and its range is 12 miles (20km).

This Flogger G (ex '488') was especially painted for the JG 9 disbandment ceremony on December 31, 1990. Together with all other MiG-23MLs, MFs and UBs it is kept in storage at Laage.

The MiG-23MF was in NVA service since 1978. It is 3,300 lb (1,500kg) heavier than the -ML version.

A MiG-23MF 'Flogger B' during take-off at Peenemünde.

A MiG-23MF in Luftwaffe colours. The old serialnumber '585' can still be read on the centerline fueltank.

Immediately after landing the wings of the MiG-23 are swept back to improve taxiing and parking performance.

The MiG-23UB is the basic trainer version of the MiG-23.

One of JG 9's five trainers. JBG 37 had four trainers of which one ('101') has crashed.

A MiG-23UB lands at its home base Drewitz, only a few miles from the Polish border.

'107' starts its take-off run from Peenemünde.

MiG-23UB '102' returns from a weather reconnaissance flight, the traditional start of a flying day.

This photo shows the functioning of the two periscope mirrors. Also note the tightness of the cockpit space, especially for the rear pilot.

Ex '109' of JG 9 in storage at Laage. The Luftwaffe is trying to sell all MiG-23s, apart from some examples that went to test facilities in several western countries.

During the summer of 1990 many East German Geschwadern adopted unit insignias. This one represents JG 9 at Peenemünde.

Night mission from Drewitz. This official NVA photographs shows how the East Germans tried to confuse western intelligence services by painting fake serialnumbers on their aircraft. Probably the pictured MiG-23BN is the '705' from the second squadron of JBG 37. (Zentrale Bildstelle der NVA-LSK/LV)

A flight of Flogger Hs from Drewitz. For deception reasons the aircraft serialnumbers on this official NVA photograph have been changed. (Zentrale Bildstelle der NVA-LSK/LV)

MiG-21bis '893' just before it leaves the flightline at Marxwalde. The bis is a fourth generation MiG-21.

JG 8 at Marxwalde was the only unit that operated the Mig-21bis, or Type N75 as it is designated by the Mikoyan design bureau.

The primary task of the MiG-21bis was air defence. An air defence training mission never lasted longer than 45 minutes.

This basic version of the MiG-21bis was identified by the ASCC as 'Fishbed L'.

These pilots, belonging to one of the three squadrons within JG 8 finished their practising drill in the cockpit of their aircraft, prior to that day's flight operations.

Externally, the MiG-21bis distinguishes itself from its predecessors MiG-21MF and MiG-21M because it has a wider and deeper dorsal spine.

The MiG-21bis has upgraded avionics and is powered by a Tumanski R-25 engine.

The MiG-21MF (like this one from JG3 at Preschen) looks very similar to the MiG-21bis.

This MiG-21bis has extra RSBN (ILS) antennas under the air intake and on the fin tip. It is identified as MiG-21bis SAU or 'Fishbed N'.

For long-term storage, this MiG-21bis SAU was preserved at Marxwalde in July 1990.

Serialnumbers of operational aircraft were painted red, while the trainer versions carried black codes.

This photo shows some of the sensors and aerials of the MiG-21bis SAU: the pitot static boom with pitch and yaw vanes, the radar radome, angle-of-attack sensor, ILS- and IFF antenna.

A JG 8 pilot checks the pitot tube of his MiG-21bis prior to a practice flight.

Cockpit interior of a MiG-21bis SAU.

Fishbed L ex-'873' of JG 8 is equipped with a centerline ECM pod.

MiG-21bis '24-28' was the former '849'.

This MiG-21MF accumulated 1350 flying hours when it became non-active on 27 September 1990.

MiG-21MF '652' belonged to JG 1 at Holzdorf. The MiG-21MF and M both belong to the third generation Fishbeds.

MiG-21MFs equipped two squadrons of JG 1 and one squadron of JG 3. This example is a MiG-21MF-75 that has a similar cockpit as the MiG-21bis. (Uwe Sauder)

Two aircrafts of the third squadron of JG 3 during their last flight in NVA service on September 27, 1990. (Uwe Sauder)

The third squadron of JG 3 would have transitioned to MiG-29 by now if the political situation in East Germany had not changed so dramatically.

A MiG-21MF 'Fishbed J' of JG 1 'Fritz Schmenkel' that was based at Holzdorf, south of Berlin.

MiG-21MF '687' of JG 3 returns at its homebase Preschen.

The dragchutes were released at the taxiway just off the runway.

The dorsal spine of the MF version is smaller than that of the bis.

The 'black box' of the MiG-21, called SARPP-12 is also essential to realize a short turnaround time. Immediately after each flight the data is examined and the aircraft is cleared for the next flight.

MiG-21 tactical codes in the 500 and 600 series were reserved for the M and MF versions

A MiG-21MF brake parachute is loaded into its compartment.

Cockpit interior of a standard MiG-21MF, or Type 96.

To commemorate the disbandment of JG 1 on 31 December 1990 this MiG-21MF (ex '510') of the second squadron was especially decorated.

Another JG 1 aircraft awaits its destiny at Drewitz. After disbandment of the units all MiG-21s were stored here and at Rothenburg.

A Preschen-based MiG-21M with a tactical reconnaissance pod.

The MiG-21M (Type 88) was used as an interceptor at Neubrandenburg and in the reconnaissance role at Drewitz and at Preschen.

After a flight this MiG-21M of TAFS 47 is pushed back manually into the flightline at Preschen.

TAFS 47 had 13 MiG-21Ms at its disposal. The tactical reconnaissance equipment was produced locally in East Germany.

Some of the aircraft of TAFS 47 were received from JG 7 at Drewitz that was disbanded on September 1, 1989 as a result of the CFE negotiations.

This MiG-21M belonged to TAFS 87 at Drewitz. The unit was formed when JG 7 was disbanded. The JG 7 aircraft went to JG 2, TAFS 47 and TAFS 87.

A MiG-21M of JG 2, homebase Neubrandenburg. In 1987, the 50 MiG-21SPS aircraft of JG 2 were scrapped and the unit received JG 7's MiG-21Ms.

The MiG-21M instrument panel.

Upon disbandment TAFS 87 had an inventory of 12 MiG-21Ms, three of which were equipped with a small camera compartment under the right wing.

A pair of MiG-21s return to Preschen. Preschen is located so close to the Polish border that instrument approaches from the east would overfly Polish territory.

To avoid FOD the cover of the air intake suction relief door is removed just before the aircraft starts to taxi out.

The camera compartment of the MiG-21M in the reconnaissance role.

MiG-21M ex '493' of JG 2.

Spraying over the East German National insignia, Preschen.27 September 1990.

In addition to FAG 15, one squadron of both JG 1 and JG 2 were also equipped with the MiG-21SPS-K. The SPS-K has no angle-of-attack sensor and has no built-in guns.

A MiG-21SPS-K of FAG 15 'Heinz Kapelle' at Rothenburg. This unit took care of training of East German MiG-21 pilots. It ceased its flying activities on September 20, 1990.

A pilot in a high altitude pressure suit steps into his MiG-21. The MiG-21 has been in use in more than 40 countries.

Two red stars indicate the years in which this aircraft went to Astrakhan in the USSR (Soviet equivalent of 'Nellis AFB') for live missile firing.

Cockpit interior of MiG-21SPS-K. (Soviet designation MiG-21PFM, or Type 94K).

MiG-21SPS 'Fishbed J' with pilot in pressure suit. The air intake of this second generation MiG-21 is smaller than that of third and fourth generation production examples.

MiG-21SPS-K of JG 1 with special painting scheme for disbandment ceremony on 31 December 1990.

JG 1 at Holzdorf stopped its operational activity within NVA on 26 September 1990, however the unit incidentally continued flying until 31 December 1990, when it was disbanded.

Another Holzdorf based MiG-21SPS-K in Luftwaffe colours.

A crewchief signals 'his' MiG-21UM to stop at the right spot at the Preschen flightline.

The trainer version of the MiG-21M could be found in every unit that was equipped with MiG-21M/MF/bis.

A MiG-21UM 'Mongol', or Type 69 is prepared for a weather reconnaissance flight at Marxwalde.

One of the Mongols that equipped the third squadron of JG 3 at Preschen.

One of two MiG-21 trainers that were assigned to TAFS 87 at Drewitz.

MiG-21UM '266' belonged to TAFS 47 Preschen.

One of TAFS 47's MiG-21UMs was equipped with these special cameras.

Lt Col Sauder, C.O. of JG 3's third squadron prepares for his last MiG-21 flight at Preschen, 27 September 1990.

Traditionally, after a pilot made his last flight in the NVA, the pilot would have a toast with his crewchief. During the last days of September 1990, a lot of toasts like this were made........

Dark clouds over Holzdorf. MiG-21UM '264' about to taxi out for its final flight.

The basecommander of Marxwalde signs for the reception of his aircraft prior to a flight.

Lt Col Lahner, the C.O. of JG 8 boards his MiG-21UM to make a weather reconnaissance flight, being the start of the flight operations on 30 July 1990.

Coded '213' this MiG-21UM previously belonged to TAFS 87 at Drewitz.

This Mongol is a MiG-21US or Type 68. It lacks an angle-of-attack sensor. The type is derived from the MiG-21SPS.

Besides the -UM and -US twoseaters, FAG 15 at Rothenburg also used MiG-21U (Type 66) trainer aircraft which belonged to the first generation MiG-21s. The dorsal spine is very small and it has a different brake parachute compartment.

Fitters resembled the most powerful offensive aircraft of the NVA. Apart from its two built-in 30mm guns it can carry 8,818 lb (4,000 kg) of various armament on 10 hardpoints. The M-4 is also capable to carry air-to-air missiles.

Standard configuration for the reconnaissance/electronic support task: one centerline KKR-2 reconnaissance container and a SPS electronic jamming pod on an underwing pylon station.

The single seat version is designated Su-22M-4. In comparison to the M-3 it is identified by extra air intakes in the nose and at the base of the fin. This one belonged to JBG 77 'Gebhard Leberecht von Blucher' at Laage.

A 'Fitter K' in front of its hardened aircraft shelter at its home base, Laage. This aircraft can be armed with the HS-58U high-speed anti-radiation missile with a range of 25 miles (40km).

An 'electronic' configured Su-22 of MFG 28 'Paul Wieczorek', also based at Laage.

A Su-22 equipped with 250 kg practice bombs ready for take-off.

One Geschwader belonged to the Air Force, while another Geschwader with anti-shipping commitments was put under Naval command.

A Fitter with its wings swept at 63 deg. During take-off and landing the sweep of the wings is set at 30 deg.

Apart from its built-in cartridges this Su-22 carries four sets of chaff/ flare dispensers on the top of the fuselage.

With four external tanks and a bomb load of 4,410 lb (2,200kg) the combat radius of this aircraft is about 500 miles.

An East German Navy Su-22M-4 rolls to the runway at Laage on August 28, 1990.

To pay attention to the fact that the unit was no longer operational after the re-unification of the two Germanies, this JBG 77 aircraft was specially decorated.

September 27, 1990 was the day on which the last Su-22 flight operations were planned.

The Su-22 has seen only four years of active service in the NVA.

Unit insignia of JBG 77, the East German Air Force Fitter wing at Laage.

A Navy Su-22 with practice bombs and 57mm rocket launchers.

To avoid corrosion of unpainted metal parts by gunsmoke, these parts were greased prior to gun firing.

The spacious and well organized cockpit of a Su-22M-4.

Electronic jamming pod and KKR-2 reconnaissance container. In addition to cameras and side-looking airborne radar, 152 flare launchers for night photography and an electric power generator equip the KKR-2.

A couple of JBG 77 pilots just after they returned from a training flight.

After disbandment of the Fitter units, most Su-22s are kept in storage at Laage (pictured is the ex '629'), while some aircraft went to research facilities in (West) Germany, Great Britain and the USA.

The trainer version of the Fitter is known as Su-22UM-3K.

In spite of the rear cockpit also being quite spacious, the view to the front is very poor. Therefore a periscope is required for the back-seat pilot.

A Fitter G about to take-off for a gun firing practice flight, as can be concluded from the greased metal parts around the gunmount. The trainer version of the Fitter has only one 30mm gun.

A Su-22UM-3K of MFG 28. Both the naval unit and the LSK/LV unit JBG 77 each had four trainer aircrafts at its disposal.

MFG 28 Fitter '138' taxies to the runway to take off for an evening flight. The Su-22UM-3K has eight weapons stores that are suited to carry the same types of weapons as that of the singleseater.

Both versions of the Fitter are equipped with a laser range finder in the intake centrebody.

Su-22UM-3K with extended periscope. The Fitter two seater lacks the 30mm gun in the port wingroot.

The K-36 ejection seats of the Fitter are of the zero-zero type.

Nose section of a Fitter G with all its sensors and aerials.

Su-22UM-3K on final approach at Laage. Landing speed is 150 knots (280 km/h).

Ex '138' with 57mm rocket pods.

L-39ZO Albatross returns to its home base Peenemünde after an evening flight.

Night flying at Peenemünde. Albatross '187' of FDK 33 about to start its take-off run.

Bautzen was the home base of FAG 25 'Leander Ratz' that took care of the training of NVA fighter pilots.

L-39ZO of FAG 25. The unit was based at Rothenburg during 1989 and 1990 due to reconstruction works at Bautzen.

L-39ZO Albatross at the Rothenburg flightline. The three squadrons of FAG 25 were responsible for respectively basic-, advanced- and operational training.

The Albatross can be armed with Atoll air-to-air missiles, and various armament to train ground attack tactics.

Most former Warpac countries - including the USSR - use this elegant trainer of Czech origin.

Sensor to detect icing conditions. Up till now, about 2,800 Albatrosses have been built.

The well-ordered cockpit of the L-39ZO includes a simple optical visor.

FDK 33 at Peenemünde was responsible for air-to-air gunnery target-towing. It used three modified L-39ZOs to tow ordinary sleeves and two of these L-39Vs. The L-39V was able to tow the KT-04 high-speed target.

The backseat of the L-39V contains a winch with the 5,560 ft (1,700m) steel cable. The winch is powered by this pressure turbine.

The white and orange colourscheme clearly identifies this Albatross as the unarmed target-tow version of the Albatross.

VFS 14 at Strausberg was tasked with light transport and communication flights. It had eight An-2 'Colt' aircraft at its disposal.

This An-2TD was used to train paratroopers at Strausberg.

The An-2 started its career in the NVA as far back as 1956.

Strausberg, the home base of VFS 14, was also the place where the defence ministry, as well as several military headquarters were located.

The National insignia of the NVA, as painted on an An-2. Many East German Colts went into civil hands after the type was phased out.

The characteristic piston engine of the An-2. Despite its fuel-consuming properties, this aircraft attracts interest from many para clubs.

'In case of emergency open here violently' as painted on the fuselage of An-2 '457'.

Flight deck of this 40-year old biplane.

Front view of An-26 with clearly visible navigator's bulged window at the port side.

Whereas most East German An-26s were camouflaged, this example was painted light grey when it carried out food supply flights in Ethiopia during 1985.

The An-26 had served with the NVA from 1980 onwards. It has a flightcrew of four: behind the two pilots a navigator and a flight engineer are seated.

The ramp at the back is very practical for fast loading/unloading while small trucks can be transported easily.

The Curl is capable of transporting 30 fully equipped paratroopers or cargo with a net weight of 12,127 lb (5,500kg).

In fact the An-26 has three engines because the starboard engine nacelle contains an extra turbofan that drives an electric generator.

All 12 An-26s are currently in use by Luftwaffe unit LTG 65.

A Tu-134A of TG 44 'Arthur Pieck' at its homebase Marxwalde. TG 44 also used three civil registered Tu-134As.

After the re-unification Marxwalde became Neuhardenberg and TG 44 became LTG 65.

The NVA-LSK maintained a considerable VIP flight. TG 44 also operated aircraft in Interflug colours.

The previously civil registered two Tu-154s also were adopted by LTG 65.

Based at Berlin-Schönefeld, three former civil Il-62s also belong to LTG 65.

The Zlin Z43 was in use as communication aircraft by VFK 1 at Cottbus, VFK 3 at Neubrandenburg, VFS 14 at Strausberg and TAS 45 at Kamenz.

One of the 12 NVA-LSK Z43s at its homebase Strausberg.

12 L-410 'Turbolets' equipped VFS 14 and TAS 45. This type has been in use since 1981.

A L-410UVP of VFS 14 at Strausberg. Four L-410UVPs were in use as communication aircraft. The eight L-410Ts had a cargo door and were capable to transport 12 paratroopers each.

Flightdeck of the Turbolet.

The four Strausberg-based L-410UVPs remain in service with LTG 65 at Neuhardenberg.

Passengers compartment of the L-410UVP. Indeed a communications aircraft!

Mi-2s were used for the training of helicopter pilots and liaison duties.

A Mi-2 of HAG 35 'Lambert Horn' at Brandenburg-Briest that was converted to carry out medivac duties.

A Mi-2 of HSFA 5 at Basepohl. Also HSFA 3 at Cottbus used this type for observation, reconnaissance and liaison flights.

This Mi-2 carries a camera for photo-mapping.

The Mi-8T is the unarmed variant of the famous 'Hip' helicopter, of which over 8,000 are used in some 40 countries.

A Mi-8T in the SAR task. During recent years the racks to carry rocket pods were removed and the Mi-8T obtained more peaceful tasks, as a result of the CFE talks.

A Mi-8T ' Hip C' on alert at Basepohl. The type was in use within the NVA since 1968.

Mi-8T '133' of KHG 5 takes off for a weather flight. KHG 5 at Basepohl was attached to the fifth army district with its HQ at Neubrandenburg.

Before conversion to a SAR helicopter, this Mi-8T was capable of carrying four pods with 16 S-5 rockets (57mm) each.

The Mi-8TB is an armed version of the Hip and came into use within the NVA in 1977.

It equipped KHG 3 at Cottbus and KHG 5 at Basepohl. This one belonged to the third squadron of KHG 5 that had 12 helicopters of this type in its inventory.

A Hip C of KHG 3 at Cottbus. This unit was assigned to the southern army group.

A Cottbus based Mi-8TB prepares to take off. KHG 3 at Cottbus operated in conjunction with army district 3 that had its headquarters at Leipzig.

The Mi-8TB is standard equipped with a 12.7mm machinegun that is operated by the flight engineer.

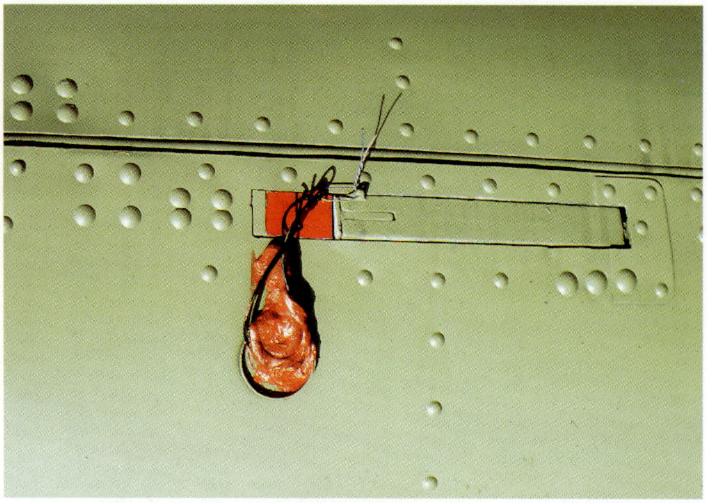

In the East German military community every door had a lock and a seal including entrances to aircraft. This photo shows the seal on a Mi-8 entrance door.

On each side the Hip C is able to carry three pods for 32 S-5 rockets each, or a 250 kg bomb. On top of the rack three launchers for AT-2 'Swatter' anti-tank missiles can be found.

The Mi-8S is a passenger version of the Hip. It can be identified by square windows. This one belonged to TG 44 at Marxwalde.

LTG 65 took over six Mi-8S helicopters in the VIP transport configuration. They are based at Neuhardenberg.

A Mi-8S of HSFA 5, Basepohl. This aircraft was configured to carry 11 passengers. (Harry Berger)

Another Mi-8S, this time in the colours of naval unit MHG 18 'Kurt Barthel' is about to land at its homebase Parow.

The naval tasks for the Mi-8TB were: detect and combat surface targets at sea and on the coast. MHG 18 was commanded by the Navy Command in Rostock.

The Volksmarine Mi-8TBs could take the same armament as the anti-tank version: machine gun, guided and unguided rockets.

The Mi-8T was in principal used for general transport duties, while some were converted to perform SAR tasks over sea.

The Mi-9s are suited to serve as a mobile command post for army commanders. Together with Mi-2s and Mi-8s they equipped HSFA 3 and HSFA 5 that were allocated to respectively the army command south and -north.

Externally, the Mi-9 differs from the Mi-8 family by some specific antennas under its tailboom.

In general, Mi-9s carried a lighter camouflage pattern than the Mi-8TBs. This one has a two tone green pattern.

The Mi-14 'Haze' is a Mi-8 development that is capable to land on water.

The NVA operated two versions of the Mi-14. This photo shows a Mi-14PL which was tasked with anti-submarine-warfare.

Sensors of the 'Haze A' included sonobuoys, a dipping sonar and MAD equipment. The weapons load could consist of torpedoes or depth charges.

This MAD 'bird' is towed by a cable through the water. It detects disturbances in the earth magnetic field which may be caused by a submerged submarine.

Entrance door and hoist of a Mi-14PL. The technical crew of this aircraft ('625') earned two quality stars for maintenance state.

Another Haze version is the Mi-14BT. Since 1989 it was in use for SAR tasks only.

During 1989 the Mi-14BTs were converted to SAR platforms.

Originally built as minecountermeasures helicopter, the 'Haze B' operated in combination with a surface ship to clear seaways from mines.

Externally the Haze B differs from the 'A' by having no anti-submarine-warfare gear and by a fuselage strake and a pod on the starboard side of the fuselage.

The bulged windows in the rear are now in use for observation.

A Mi-14BT in SAR configuration on alert at Peenemünde.

Like the MiG-29, the Mi-14 is also equipped with an audio warning system that informs the pilot when the aircraft's safety limitations are exceeded.

The most effective battlefield helicopter of the NVA was the Mi-24 'Hind'. This photo shows one of the 20 aircraft that equipped the first squadron of KHG 5 at Basepohl.

The Mi-24D 'Hind D' is standard equipped with a 12.7mm Gatling gun. It is controlled by the weapon operator in the front seat. The pilot occupies the back seat while an engineer finds a place between the cockpit and the main cabin.

A Mi-24D of KHG3. The type was in use by the NVA since 1979. In the assault role, the Hind would typically carry eight fully armed airborne troops.

A Hind D with UB 32 rocket pods is seen taking off from Cottbus. Maximum armament load is 3,970 lb (1,800kg).

The crew is secured by heavily armoured plates around the cockpits which have bullet-proof windscreens.

The starboard undernose pod incorporates FLIR and LLTV equipment while the port one contains a missile guidance antenna.

In addition to the four AT-2 'Swatter C' guided anti-tank missiles the Hind D is capable of carrying four UB-32 pods or a bombload of 2,205 lb (1,000 kg). The wingtip contains a gun camera.

After the lessons learned in Afghanistan, all Hind Ds were retrofitted with these chaff/flare dispensers. Also IR jammers could be installed behind the rotor shaft.

A pair of KHG 5 Hind Ds during their last flight in NVA colours on September 26, 1990. (Hendrik Jan van Broekhuizen)

The Mi-24P or 'Hind F' is a more sophisticated version of the famous gunship. Major external differences are the twin-barrel 30mm gun, modified wingtip launchers and chaff/flare dispensers that moved forward, just after the national insignia.

Mi-24P '444' of KHG 5 prepares for a dusk flight. It carries a pod for twenty 80mm unguided missiles. The Mi-24P is also capable to take a load of 16 AT-6 'Spiral' guided anti-tank missiles.

One of the 12 Hind Fs that equipped the second squadron of KHG 5.

In addition to various flight instrumentation, navigation and communication equipment, the Hind incorporates IFF, radar warning antennas and an optional infra-red jammer.

A Basepohl-based Hind F over East German farmland. (Hendrik Jan van Broekhuizen)

This photo shows the air data sensor pod, combined FLIR/LLTV pod, 30mm cannon and missile guidance pod.

To improve its self-defence capability the Hind F can be armed with R-60 'Aphid' infra-red homing missiles.

Both the weapon operator and the pilot have a reasonable view to the front and to the sides.

MODERN GERMAN AIRPOWER

4540

DML
PLASTIC MODEL KITS

DML BRINGS YOU THE BEST IN 1/144 SCALE AIRCRAFT MODEL KITS. EACH PLASTIC KIT FEATURES LARGE COLORFUL DECAL THAT RESEMBLES THE AUTHENTIC OR CEREMONIAL PAINTING SCHEME OF THE AIRCRAFT.

4546

4554

4555

4556

1014 **Modern Soviet Warplanes: Fighters & Interceptors**
by Steven J. Zaloga

1015 **Modern Soviet Warplanes: Strike Aircrafts & Attack Helicopters**
by Steven J. Zaloga

1017 **F-117A Stealth Fighter**
by Andy Sun

1019 **U.S. Airpower at Sea**
by Richard S. Drury

1020 **The ATF Contenders: YF-22 & YF-23**
by Andy Sun

1022 **Operation Daguet: French Air Force in the Gulf War**
by Eric Micheletti

AN ABSOLUTE MUST FOR ALL **SERIOUS** AVIATION MODELERS AND ENTHUSIASTS.

CONCORD
PUBLICATIONS COMPANY

4002 **Super Show! International Air Tattoo & Tiger Meet 1991**
by Ian Rentoul, Tom Wakeford & Mark Attrill

2004 **U.S. Airpower in Desert Storm**
by Michael Green

2005 **Gulf War: British Air Arms**
by Ian Rentoul & Tom Wakeford

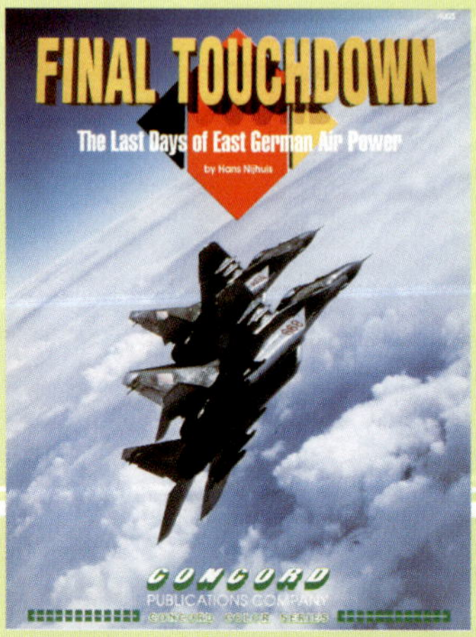

4003 **Final Touchdown: The Last Days of East German Air Power**
by Hans Nijhuis